CRANBERRY

IN A NUTSHELL

CRANBERRY

VACCINIUM MACROCARPON

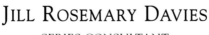

JILL ROSEMARY DAVIES

SERIES CONSULTANT

ELEMENT

SHAFTESBURY, DORSET • BOSTON, MASSACHUSETTS • MELBOURNE, VICTORIA

Series Consultant Jill Rosemary Davies
Text Jill Nice

First published in Great Britain in 2000 by
ELEMENT BOOKS LIMITED
Shaftesbury, Dorset SP7 8BP

Published in the USA in 2000 by
ELEMENT BOOKS INC.
160 North Washington Street,
Boston, MA 02114

Published in Australia in 2000 by
ELEMENT BOOKS LIMITED
and distributed by
Penguin Australia Ltd.
487 Maroondah Highway,
Ringwood, Victoria 3134

NOTE FROM THE PUBLISHER
Any information given in this book is not
intended to be taken as a replacement for
medical advice. Any person with a
condition requiring medical attention
should consult a qualified practitioner
or therapist.

For growing and harvesting, calendar
information applies only to the northern
hemisphere (US zones 5–9).

Designed and created for Element Books with
The Bridgewater Book Company Ltd.

ELEMENT BOOKS LIMITED
Editorial Director Sue Hook
Project Editor Kate John
Assistant Editor Annie Hamshaw-Thomas
Group Production Director Clare Armstrong
Production Controller Hannah Turner

THE BRIDGEWATER BOOK COMPANY
Art Director Tony Seddon
Designer Jane Lanaway
Editorial Director Fiona Biggs
Project Editor Lorraine Turner
DTP Designer Trudi Valter
Photography Guy Ryecart
Food styling Fiona Corbridge
Illustrations Michael Courtney
Picture research Liz Moore

Printed and bound in Portugal

Library of Congress Cataloging
in Publication data is available

British Library Cataloguing
in Publication data is available

ISBN 1 86204 707 3

*The publishers wish to thank the following
for the use of pictures:*
AKG London Ltd: pp.10t, 13cr,
14tr, bl, 20t, 27t;
A–Z Botanical: pp.10b, 16br, 28t,
50cr, 58;
Bruce Coleman Collection: pp.6t, b,
20bl, 53b;
Science Photo Library: pp.17, 23bla,
blb, 53t;
Tony Stone Images: pp.2, 7tl, 11b, 24b,
27b, 30, 43b, 51b.

Special thanks go to: Dina Christy for
organizing studio photography.

Contents

Introduction

ABOVE
*Cranberries are
rich in vitamin C
and minerals.*

LONG BEFORE THE PILGRIM FATHERS *discovered the medicinal and culinary benefits of Cranberries, native peoples in both Northern Europe and the, as yet, undiscovered New World were knowledgeable about the therapeutic qualities of all members of the Vaccinium family, which includes Blueberries, Bilberries, Cowberries, and Cranberries, and used them to remedy ailments as diverse as scurvy and gout.*

The Cranberry plant, growing in the wild, is a neat, evergreen, prostrate shrub. The small oblong-ovate leaves – similar in appearance to box – are shiny, dark-green above and gray-green underneath, with slightly curling edges. When the plant blooms in midsummer, it has delicate, rose-tinted flowers, which grow singly or in groups of as many as ten on a single shoot upon slender stems arising from the leaf axils. The unexpanded flower and slender corolla resemble a crane's head and neck – hence the name Cranberry. The fruit is ready for picking in early fall.

Cranberries are invaluable sources of vitamin C, minerals, and fiber – they are also strongly astringent with a high acid content and little natural sugar.

RIGHT
*Cranberries
are rich in
vitamin C.*

ABOVE *Peat bogs and marshland are the native habitat of the Cranberry plant.*

helps to preserve them, so they are slow to decay and dry well. The American Cranberry is more robust than the European variety, but will thrive only on heathland. Its preferred habitat is the acidic soil and peat of open bogs, wet shores, and grassy swamps, and, occasionally, poorly drained upland meadows and salt marshes.

Cranberries are chamaephyte shrubs, meaning that their winter buds are situated close to the soil's surface.

Blueberries

The scarlet berries are round, becoming slightly oval at one end, and are usually larger than blackcurrants, which they resemble. The larger American variety, *Vaccinium macrocarpon*, is particularly rich in pectin, which means it is excellent for making preserves and requires little cooking, so loss of nutrients is minimal. This is one reason why the Cranberry became such a useful fruit for explorers and settlers. In the past, it was popular in juices, jellies, and relishes, which enlivened the table and provided a much-needed source of vitamins throughout the months when fresh fruit was unavailable.

The fresh berries are covered with a waxy coating, which

DEFINITION

Botanical family: Cranberry is part of *Ericaceae* – the heather family. It is related to the rhododendron *Arbutus* (strawberry tree), and to other medicinal herbs such as Uva Ursi.

Species: The genus contains about 150 species, and because of the confusion arising from the intermingling of botanical and common names within folklore, identification is somewhat erratic. However, the best-known species, the Large or American Cranberry (*Vaccinium macrocarpon*), has been developed by selective cultivation on a commercial scale from the indigenous Cranberry (*Vaccinium oxycoccus*) of North and Central Europe, North Asia, and North America.

Cranberries

Exploring Cranberry

RECORDS SHOW THAT *early European adventurers on the American continent could survive on a bag of dried Cranberries alone, indicating that they were aware of the berries' nutritional and healing powers. Cranberries have an ancient history but are best known as a traditional part of the Thanksgiving meal.*

LEFT **Dried Cranberries were ground up for use as a medicine.**

There is, however, very little documented early history on the therapeutic use of Cranberries. There are recorded instances of the seeds of members of the *Vaccinium* family being found on Iron Age sites. Later, the Romans in Britain thought that Cranberries may have been used in primitive pagan rituals practiced by the indigenous island tribes. They also believed that these berries enabled their enemies to see in the dark. The skin of Bilberries does contain anthocyanidins, bioflavonoids, and vitamin C, which act on the retina of the eye to improve vision, particularly at night. However, in the case of the Roman invaders, these claims may have been an excuse, on occasion, for being outdone.

Throughout Northern and Central Europe, Cranberries and other members of the *Vaccinium* family were used in strengthening brews and liquors. The berries were gathered and dried to

provide both sustenance and protection against vitamin deficiency. They would have been particularly valuable in cold climates, where people had to endure long winter months without fresh food and little sun.

Native Americans had long understood the value of the Cranberry. They used it as a healing plant, and ground into "pemmican" – a mash of meat and fat – to sustain them on long hunts. A similar type of pemmican was made by the hunter-gatherers of Northern Europe and Asia. This nutritious mash was invaluable in the winter when food was scarce. However, by the time that the New World was

ABOVE *Cranberry sauce is a part of the Thanksgiving meal.*

"discovered," the Europeans had lost many of these ancient recipes.

Whether or not it is true, it is said that the Cranberry was reinstated as a valuable food on the first Thanksgiving Day. At the onset of winter, inexperienced settlers who were new to North America found themselves with failed crops and facing starvation. A tribe of Native Americans, realizing their plight, brought them meat, fruit, and vegetables, among which were turkey and Cranberries. Everyone sat down together to enjoy the feast in friendship, and the settlers gave thanks to God for their deliverance from certain death.

LEFT *The Cranberry helped to forge a friendship between the Native Americans and the white settlers.*

So highly prized did Cranberries become that, in 1677, ten barrels were shipped back to England from the American colonies as a placatory gift for King Charles II, and later they were exported to Europe on a regular basis.

ABOVE *King Charles II received Cranberries from the Colonies.*

In 1775, American officer Colonel James Smith wrote that he had seen Cranberries growing in swamps. He observed that they were gathered by Native Americans, a practice later enthusiastically adopted by colonial settlers.

In 1800, plantation owner Eli Howes produced the first cultivars from wild plants at East Dennis, Massachusetts, and, in 1816, Henry Hill, a Cranberry farmer

BELOW *As Cranberry's popularity increased, limits on picking were imposed.*

on Cape Cod, observed that his berries grew larger and juicier when sand from nearby dunes blew over the vines. From then on, these became the preferred conditions for growing berries. As cultivation took off, swampy wastelands were drained and turned into plantations. By 1912, 26,300 acres of Cranberries were farmed and 512,000 barrels of fruit harvested. The huge commercial Cranberry enterprise now harvests in excess of 4.7 million barrels of fruit a year.

As settlements expanded into towns, there was fierce competition between families to be the first into the bogs to gather sufficient quantities of Cranberries to see them through the winter months. In order to establish fair play, a rule was imposed: no one could pick more than one quart per person before September 20th each year. In 1773, in one town in Cape Cod, anyone who "jumped the gun" had the illegal haul confiscated and was fined one dollar.

COMMERCIAL GROWERS

The main commercial growers are in the United States, where Cranberries are cultivated extensively in Massachusetts, New Jersey, Wisconsin, Oregon, and Washington, as well as in Quebec and British Columbia in Canada. Every acre of Cranberry bog is supported by four to ten acres of wetlands, woodlands, and uplands. North American growers preserve almost 200,000 acres of support land, growing Cranberries on 35,000 acres. In this way, not only does commerce thrive but also vast tracts of natural habitat are provided for wildlife. Growers belong to cooperatives that enforce high standards and quality of Cranberry products.

ABOVE *Cranberries are grown commercially in the areas shaded red.*

SOIL REQUIREMENTS

The needs of the Cranberry plant are very particular. In the wild, they will only grow successfully in open bogs and marshes that contain a high percentage of sand that has an acid pH of 4.0–6.1. Under cultivation, the vine cuttings are planted in bogs or marshes where former peat swamps have been cleared and leveled before being covered with a 3in (8cm) layer of sand. A combination of acid peat soil and sand, together with a good supply of fresh water, is essential.

LEFT *Cranberries are cultivated commercially in swamp lands.*

A history of healing

ONE OF THE FIRST MEDICAL *texts to mention Cranberries was written in 1578 when Henry Lyte, a herbalist squire of Lytescarie, mentioned their use in his Niewe Herball, a work translated from the Flemish. This impressive volume can still be viewed at the National Trust property of Lytes Cary in Somerset, England.*

ABOVE **Cranberries have a long history of healing.**

Before the 17th century and the "discovery" of the New World, the therapeutic uses of Cranberry had lapsed into obscurity. This decline in popularity may have been because Cranberry was a bog-dwelling plant, and thus not as easily gathered as other species of *Vaccinium*. Another possibility may be that it was very localized and became even more so as swampy marshlands were enclosed or drained.

Both Virgil and Pliny referred to the generic name for Cranberry (*Vaccinium*) in their writings. Some people believed the word to be a corruption of "hyacinthus,"

while others considered that it was named after the cow (*vacca* in Latin). This learned dithering may be the reason

RIGHT **Versatile Cranberry is used in creams, juices, and oils.**

ABOVE *Ruby Cranberry juice is a traditional remedy for eczema.*

why there are few references to Cranberry in old herbals. There is undoubtedly confusion regarding the plant's classification – caused by the free exchange of the same common name between various species. Geoffrey Grigson, in *The Englishman's Flora*, gives some indication of how local names changed as they jumped county borders, and how one person's "Cranberry" became another's "Fenberry."

Although it is not possible to identify the Cranberry positively in ancient writings, *Vaccinium* was used in healing from the 16th century onward. Several 17th-century remedies refer to a decoction of Cranberry leaves for the treatment of gout and rheumatism, but herbal texts disagree as to whether Cranberries were an excellent cure for diarrhea or whether they relieved constipation. However, most texts concluded that the dried berry, chewed well, was generally beneficial.

The great majority of the healing remedies that continued to be known in Europe were those of the country "goodwife" (housewife), which were passed on by word of mouth. Domestic they may have been, but remedies such as mash of

ABOVE *The early settlers learned of the value of Cranberries from Native Americans.*

Cranberries and buttermilk to treat erysipelas (a bacterial infection of the skin) would probably have brought immediate and effective relief. Cranberry juice was also traditionally used to cure eczema and other skin disorders.

The Native Americans had a thorough knowledge of how to treat deficiency diseases – of which, at the time, Europeans were in relative ignorance. In 1638, John Josselyn, gentleman and traveler, observed New England natives used Cranberry to treat scurvy and fevers.

It was also discovered that some American tribes used a Cranberry poultice to draw poison from wounds, and the fruit's juice to alleviate skin rashes caused by insect stings and plants. Another

BELOW *Native American women ate Cranberries to relieve infections.*

ABOVE *Barrels of Cranberries were shipped to European ports.*

BELOW *Limes, like Cranberries, are rich in vitamin C and prevent scurvy.*

habit observed was the drinking of Cranberry juice by Native American women. This was a practice eagerly taken up by colonial settlers, who realized that it brought considerable relief for "women's troubles" – namely cystitis and urinary and genital infections.

RECENT HISTORY

A traditional cure for scurvy, dehydrated Cranberries were eaten by US troops serving abroad in World War II. Soon scientists and doctors came to realize that, in addition to its high levels of vitamin C, Cranberry had other beneficial effects – particularly in reducing

bacterial infections of the bladder. Although a controversial issue, since no scientific studies had yet been carried out, it was suggested in 1914 that Cranberries were particularly rich in benzoic acid, which is useful in the treatment of urinary tract infections. However, these findings could not be completely substantiated. From 1920 to 1970, further research was carried out and a compound was isolated (quinic acid), which is believed to be an inhibitor of bacterial growth in the urinary tract.

Drinking a high concentration of Cranberry juice was already known to prevent infections in people prone to urinary tract problems and is now widely used in their treatment. Extensive research programs have been carried out in the US at the Brigham and Women's Hospital, Utah, and Harvard Medical School, into the effects of Cranberry juice on urinary tract infections. These have shown that there is good scientific evidence to prove that Cranberry juice is an effective treatment for

RIGHT *X-rays help doctors ascertain any damage in the urinary tract.*

these conditions. Further research continues in order to isolate the specific factor or factors that have this effect. Today, Cranberry – particularly in the United States – is most obviously identified as the traditional gastronomic partner to turkey. Long used in Europe as an accompaniment to rich foods – most notably, game and venison – it plays its part not only in enhancing flavors but also as an aid to digestion.

BELOW *Cranberry juice is excellent for relieving cystitis.*

Anatomy of Cranberry

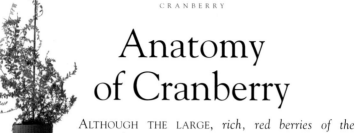

ABOVE *Cranberries and their leaves have healing qualities.*

ALTHOUGH THE LARGE, *rich, red berries of the American Cranberry are the variety that are most easily available, the small, wild berries also contain the same nutrients – simply a little less pectin. The leaves of all varieties of* Vaccinium, *except the cultivated Cranberry, have the potential for healing.*

FRUIT

The fruit of the Cranberry is the part of the plant most often used in therapeutic practice. In the fall, the low-growing shrub produces bright red, broadly oval berries similar to blackcurrants but twice the size. The waxy skins enclose a fibrous, acidic, and astringent pulp with numerous seeds, which is not entirely palatable when eaten raw. Traditional remedies generally used the dried or fresh fruit, but most modern recipes are based on the juice. The pulp and seeds of the berries are a rich source of pectin, which, when mixed with a sweetener and the natural acid of the fruit, forms a jelly.

Fruits that are gathered for juices or sauces are harvested "wet," while berries destined to be sold whole, either fresh or dried, are harvested "dry" (see page 30).

BELOW *Delicate Cranberry flowers may be seen on marshy heaths in the spring.*

Chemical constituents

Cranberries are astringent, antiseptic, and tonic. They have a high nutritional value and contain pectin, sugars, tannins (arbutin), flavonoids (anthocyanin pigments), carotene, and an extraordinarily rich mix of vitamins and minerals. Most important is the antiscorbutic vitamin C (see page 56), which is present in large quantities, and also vitamins A and B, riboflavin, and niacin. Added to the minerals – sodium, potassium, calcium, magnesium, phosphorus, copper, sulfur, iron, and iodine – is a unique blend of organic acids: quinic, malic, and citric acid. Quinic acid is at present considered to be the most important of the three.

It is the presence of tannins, however, that makes the use of Cranberry so effective in the treatment of bladder infections. Tannins also act as strong antimicrobials and help to "dry" loose bowels. They are complex mixtures of aromatic acids that

ABOVE *Cranberries have four air pockets.*

have astringent and antiseptic properties. They cause proteins to clump together in mucous membranes. This makes them more rigid and so deprives bacteria of nutrition, aiding the swift healing of wounds, reducing sensitivity and pain, and forming a protective crust. Tannins also help to dry out tissue, which prevents further infection.

Research by Howell and others in 1998 (see page 58) revealed that proanthocyanidins (see page 57) are the key to the effectiveness of Cranberry in the treatment of urinary tract infections. They prevent the adherence of

RESEARCH

Initial research into the effect of Cranberry juice on bladder infections was carried out as early as 1914. It was thought that benzoic acid present in Cranberries raised the acid level in urine and reduced infection. Further studies from the 1920s to the 1970s suggested that acidification of the urine through the drinking of Cranberry juice produced a bacteriostatic effect (see pages 52 and 56), but these reports were conflicting and inconclusive.

LEFT *Flavonoids in Cranberries strengthen blood capillary walls.*

bacteria to the mucosal walls of the urinary tract, thus keeping the bacteria from multiplying and infecting the host tissue.

Proanthocyanidins and anthocyanins are the flavonoid glycosides that comprise plant pigment – the purples and reds of autumn leaves, the mauves, reds, and blues of flowers, and the blues and reds of berries, whose color is dependent upon the pH of the soil in which the plants grow.

The flavonoids in Cranberry enable vitamin C to be better absorbed and utilized by the body. They also strengthen blood vessel walls and

increase blood supply to the heart. Flavonoids are compounds that have characteristic variants with differing activities and abilities. For example, they can be antibacterial, anti-inflammatory, antioxidant, and antimicrobial.

Cranberry capsules

Cranberry juice that is to be used for therapeutic purposes should contain at least 27% of pure Cranberry juice. Some Cranberry juices are mixed with other fruit juices to make them more palatable. They also often have water and sugar added to them. The level of sugar in these Cranberry juices concerns doctors, because it can exacerbate some medical conditions.

If you wish to avoid the added sugar, you can buy untreated, dried Cranberries from herbalists and health food stores. They are naturally sweet and pleasantly sharp.

RIGHT *Raspberries and redcurrants help to sweeten Cranberry juice.*

LEAVES

The small, shrubby leaves of the wild Cranberry plant can be used medicinally, but they are not as effective as those of other members of the *Vaccinium* family, such as blueberries or bilberries.

Although in no great strength, the leaves contain tannins, organic acids, sugars, vitamin C, and glycosides (arbutins). These

LEFT *Drinking leaf infusion may help relieve eczema.*

SHELF LIFE OF LEAVES
Fresh leaves will keep on the twig, in water, until they wilt, and in a polythene bag in the refrigerator for up to one week; dried leaves will keep for several months if stored in a dry, airtight container.

give them astringent, antiseptic, disinfectant, diuretic, and mildly hypoglycemic (lowering blood sugar) properties. An infusion or decoction can be used to treat diarrhea, bed-wetting in children, and skin diseases.

SHELF LIFE OF FRUIT
Most freshly picked or store-bought Cranberries will last at least three months, and maybe longer if kept covered and refrigerated; dried berries will stay in optimum condition for a year or so. Freezing whole berries or the pulp as soon as possible after picking is an acceptable method of preserving Cranberries without loss of nutrients. They will last up to one year in the freezer.

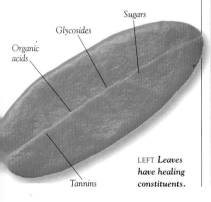

Sugars

Glycosides

Organic acids

Tannins

LEFT **Leaves have healing constituents.**

Cranberry in action

THE LEAVES AND BERRIES *of the Cranberry have always enjoyed an almost mystical reputation in Europe. A tincture was often recommended in order to retrieve the minds of people considered to be bewitched. Cranberries were scarce and their habitat somewhat remote, and the plants were difficult to harvest, so when used, their success was treated with a certain amount of superstitious awe.*

ABOVE
Cranberry was used as protection from harmful witchcraft.

ABOVE **Cranberries grow best in boggy, acidic soil.**

HOW CRANBERRY CAN HELP

※ Effective against skin conditions and maintains a healthy skin. Cranberries contain antioxidants and anthocyanidins, which help to keep the skin firm and improve elasticity, thereby helping to combat the effects of ageing.

※ Helpful in the treatment of stomach troubles such as diarrhea, enteritis, and dysentery due to the fresh and dried fruit and juice containing powerful

LEFT *Whizz up your own Cranberry drink by juicing the berries in a liquidizer.*

astringent and disinfectant properties in the form of anthocyanins and tannins; they also restore depleted minerals.

🌱 Fights colds and chills. A handful of dried cranberries daily is a natural deterrent to catching colds, particularly for children, who may find them a palatable alternative to sweets. A glass of Cranberry juice daily, with or without added fruit juice, will increase the body's reserves of vitamin C and thereby give additional protection against colds and influenza. Cranberry juice not only helps to restore health, it is also effective in reducing fever.

🌱 Stimulates the appetite and encourages the digestive juices to break down rich and fatty foods, which is the reason why Cranberry sauce is still served with rich meat and game.

🌱 Eases infections of the urino-genital tract (particularly cystitis), which result in a frequent desire to urinate, painful urination, and lower back pain. The pure concentrated juice has been shown to have very positive results in the treatment of these painful conditions.

RIGHT *A glass of Cranberry juice a day protects against colds and influenza.*

LEFT *Cranberries stimulate the appetite and help the body to break down fatty foods.*

21

HOW CRANBERRY AFFECTS THE BODY

Half a cup of fresh Cranberries provides 10% of the body's daily requirement of vitamin C. The body neither stores nor makes this vitamin; a continuous supply must be provided through daily food intake to ensure normal body cell functioning and the formation of healthy collagen, bones, teeth, cartilage, skin, and capillary walls.

Fresh Cranberries

Vitamin C helps from within to heal wounds and burns, because it assists in the formation of strong connective tissue.

Vitamin C also helps the body to utilize other nutrients – most particularly iron, calcium, amino acids, and vitamins A, B, and E. High stress levels, excess alcohol and smoking, infections, and fevers all raise the body's need for vitamin C. This vitamin stimulates the effective action of white cells and antibodies, thus strengthening the immune system, and acts as an anti-oxidant, helping to protect the body from free-radical damage (see page 56).

The astringent properties of Cranberries can shrink and firm living tissue. This reduces or stops the function of certain body tissues, with the consequence that Cranberries can bring relief from diarrhea.

Cranberries are also a natural remedy for intestinal infections. The fresh or dried berries have the advantage of passing through the digestive system without affecting it, and only begin to work when they reach the small intestine. Recent investigation would suggest that, in a decoction of berries, the disinfectant properties of members of the *Vaccinium* family are so powerful that they can sterilize bacteria in the colon.

RIGHT
Vitamin C guards the body against damage by free radicals.

Cranberry juice has a tonic effect on the body and purifies the blood. Certain compounds in the fruit help its nutrients to be assimilated into the bloodstream quickly, thus fortifying the body.

Reports from the Journal of American Medical Association in 1994 indicate that ongoing research could show Cranberry juice produces impressive results in the treatment of urinary tract infections such as kidney and bladder inflammation, painful urination (cystitis), and kidney stones. Cranberries contain acids that are not oxidized in the body. They are rich in quinic acid, which increases urine acidity and decreases the levels of alkali, urea, and uric acid without causing acidosis. By

RIGHT *Cranberry juice reduces bacterial infections.*

lowering the pH, Cranberry juice creates a less favorable environment for bacteria in the intestines and genitalia. It also appears to stop bacteria from adhering to mucosal surfaces, thus preventing infections of the bladder and urethra (see pages 17–18).

LEFT *Bladder infections, and kidney stones (below), respond to Cranberry treatment.*

WHEN TO AVOID CRANBERRY

Although unlikely, it is possible that anyone drinking 4 cups (1 liter) of Cranberry juice daily over a long period of time could develop kidney stones, due to Cranberry's acid content. Paradoxically, up to 10fl oz (300ml) Cranberry juice daily *reduces* the risk of kidney stones and helps to prevent urinary tract infections. Irritable bowel syndrome sufferers may find that copious amounts of Cranberry fruit and juice, and little else to eat, may aggravate the condition and cause diarrhea.

Energy and emotion

Dried Cranberries

THE TASTE OF CRANBERRY JUICE *and of the fresh or dried berry is astringent and acidic. It leaves the mouth feeling thoroughly cleansed and it is this characteristic that indicates its effect upon the body.*

Cranberry is a tonic plant, and it is the qualities of such plants that cleanse the body, ridding it of impurities. This is particularly true of the effect that Cranberry has upon the kidneys and urinary tract. When this system malfunctions, it can create feelings of overwhelming lethargy, combined with biliousness and headaches, which, in turn, lead to depression and a sense of feeling "under par." Constipation can also provoke this same listlessness and, once again, the unique way in which the fresh or dried Cranberry fruit acts upon the intestinal and digestive systems promotes an

ABOVE *Boost recovery from illness with a Cranberry tonic.*

effective remedy to regulate them and dissipate toxicity.

Due to the catalyzing effect of certain mineral salts in Cranberries, the chemistry of the blood is purified. This immediately has an uplifting and invigorating effect upon the body, mind, and emotions.

When the body is under siege from illnesses such as colds and flu, or from injury or self-induced ailments related to stress (for example working too hard, smoking, excessive coffee intake, lack of exercise, and inadequate diet), a large amount of vitamin C is required to restore the body's ability to function properly.

Without an adequate supply of fresh fruit and vegetables, particularly during the winter months, the body could lack vitamin C, and the Cranberry fruit immediately replenishes it, healing the system from within and restoring depleted energy levels. Thus, tension is relieved, pain is eased, and a sense of well-being becomes apparent.

When the balance that allows the organs of the body to function correctly is achieved, emotional and physical harmony are restored. Cranberry ensures that the body's process of assimilation and elimination are at their most effective.

ENERGY AND THE MIND

Wise people know that heartache and unhappiness are only made worse by dwelling on misery. In the past, Cranberry was used as a tonic "to draw out dark thoughts" and so it became known as "the balm to an aching heart."

Traditionally, all members of the *Vaccinium* family were used to counteract the effects of the "black arts." Cranberry fruit was used to "purify people" of the deep fears, darkness, and isolation engulfing those who had been involved with the "black arts in this and past lifetimes."

As a tonic plant, Cranberry can, through its physical effects upon the body, have a profound and positive effect on the mind. It flushes out toxins and, as a result, releases poisons from the mind, body, and spirit. When the body is feeling more energized, the mind becomes less introspective. When energy is renewed, life can be approached with restored vitality and enthusiasm.

ABOVE
Cranberry juice gives confidence and optimism.

RIGHT **A constant supply of vitamin C is needed to maintain health and vitality.**

FLOWER REMEDIES

Cranberry flowers cling with tenacity to the swampland vines on which they grow. They imbue the spirit with a sense of survival and wisdom. This flower remedy is therefore all about helping to fight fear and reduce vulnerability. It helps to calm people's fears, whether real or imaginary, thereby soothing the mind as well as the body.

TO MAKE A FLOWER ESSENCE

STANDARD QUANTITY

Approx. 1½ cups (375ml) each of spring water and brandy, and 3–4 Cranberry flowers.

2 Use a twig to remove the flowers from the bowl. Measure the remaining liquid and add an equal measure of brandy to preserve it.

3 Finally, pour the mixture into sterilized, dark glass bottles and label them.

Recommended dosage

Adults: 4 drops under the tongue 4 times daily, or every ½ hour in times of crisis. Children: over 12 years, adult dose; 7–12 years, ½ adult dose; 1–7 years, ¼ adult dose; younger than 1 year, consult a herbalist.

1 Submerge carefully chosen Cranberry buds and flowerheads – freshly picked in the early morning – in a glass bowl of shallow spring water. Cover the bowl with clean, white cheesecloth and put in the sunshine for at least three hours. Try to ensure you have three hours of continuous sun. If the flowers wilt sooner than this (they may, depending on the strength of the sun), they can be removed earlier.

LEFT *In the United States, Cranberry became a symbol of tenacious survival in the face of hostile conditions.*

PLANT SPIRIT ENERGIES

In Europe, the spirit of the small and secretive Cranberry plant was considered to be one of humble yet persistent optimism. This led to people using the leaves and fruit to treat and relieve symptoms that were causing depressive states.

However, in the New World, the Cranberry became a symbol of survival and strength. The very nature of its exclusive habitat imparted a sense of tenacious endurance in the communities nearby. This plant can be judged by the effects that its bright berry has upon mind and spirit, and this must be allied to its physical effects, which are, in essence, those of clarity and a bracing, healthy, robustness. Perpetually wholesome and invigorating, it did, at one time, ensure the survival of Native Americans and, due to the generosity of those people, brought growth and renewal to another society in need. These are the qualities that Cranberry continues to share with us.

RIGHT *People who seize life with both hands mirror the spirit of Cranberry.*

Growing, harvesting, and processing

ABOVE **Berries follow pale pink blossoms in June.**

CRANBERRIES WERE ONCE *and are still known as "The Ruby of the Bogs" and "Red Gold" because of their tendency to grow on vines in bogs or marshes. At the end of the 19th century, demand in the UK was such that Cranberries were imported from Russia and Scandinavia; during World War II tons of berries were sent to American troops serving abroad.*

GROWING CRANBERRY

Commercial Cuttings are taken in middle or late spring from well-established cultivars and set into peaty soil that has been covered with a layer of sand. Fresh water is crucial for successful cultivation. This is controlled by a system of dikes, ditches, and drains.

Cranberry roots are fine and fibrous, with no root hairs, and spread throughout the top 3in (8cm) of soil. The root zone of the plant must be moist but not saturated. The bogs are kept dry during the summer and wet during the rest of the year. If frost

ABOVE **European Cranberries growing in an acid bog.**

threatens as harvest time approaches, or the vines need protection during extreme weather, Cranberry plantations are flooded with fresh water overnight to protect the crop.

An efficiently cultivated Cranberry bog will bear fruit three to five years after the vine cuttings are set and will continue to produce berries indefinitely. Pruning the vines takes place in the spring or fall and weeding in the summer. Fertilizing and resanding is done every three to four years to maintain the bog.

Homegrown To grow Cranberries successfully, a moist, peaty soil is required with no lime present. Dig a trench 2½ft (75cm) deep. Cover the base with medium-size stones and a layer of peat and leaf mold. A further layer of sand – 3in (7.5cm) deep – will prevent the peat from drifting when flooded and help to retain moisture. Plants should be set 2ft (60cm) apart.

Planting should take place in the early fall, and if it is not possible to

obtain locally grown cuttings, then buy the American Cranberry (*Vaccinium macrocarpon*), which is the most suitable for cultivation. Immediately after planting, the whole bed should be flooded with rainwater or chalk-free water. Do not water again until the following summer (if it is a very warm spring, it may be necessary to water earlier), when the bed should be flooded from time to time. The rooting medium should never be allowed to dry out completely. Provided that conditions are as close as possible to the Cranberry's natural habitat, the bed should require no further attention, except for occasional weeding.

It should be noted that, because the American Cranberry is a cultivar that has been selected and bred for the excellence of its fruit, its leaves do not have the same therapeutic qualities as those of the wild Cranberry.

RIGHT **Established plants can be propagated by division.**

HARVESTING

Commercial By the fall, it is time to begin the harvest. At one time, pickers would work on their knees, pushing scoops in front of them as they moved through the bogs. The scoops had tines, and were large enough to go across a person's knees. Good, firm berries would pass through the tines without damage. A skilled scooper could harvest an average of 100lb (45kg) of Cranberries an hour. In areas where it is impossible to use mechanical pickers, this method of harvesting is still employed.

Nowadays, mechanical pickers, which look like large lawnmowers, are guided around the bogs where they gently comb berries from the vines with moving metal teeth. A conveyor belt then carries the fruit to a box or bag at the rear of the machine. This method – called "dry" harvesting – produces berries that are subsequently sold fresh or dried.

Fruit for processing into juice or sauces is "wet" harvested. The bogs are flooded to a depth of 18in (46cm), then a "water reel" machine, similar to a giant eggbeater, churns the water to dislodge the berries. In this way, the berries float up to bob on the surface of the "lake." The floating berries are corralled by wooden booms called "racks" and taken to waiting trucks.

Homegrown The berries will be ripe in the early fall, when they are bright red and firm. Pick them with your fingers or use a scoop. Berries that are squashed or damaged should be discarded. Good, bouncy berries can be frozen and used at a later date, since they lose few of their valuable nutrients this way.

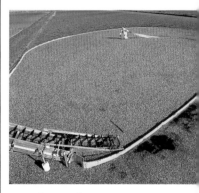

ABOVE *A spectacular sea of ripened fruit ready for harvest.*

PROCESSING

Commercial Cranberry growers deliver their berries to a receiving station where a machine blows the leaves and twigs from the fruit.

Fruit to be sold fresh must pass a strict quality test, which makes use of the fact that a good berry will bounce over mini wooden barriers. If a berry does not bounce, it is discarded. "Wet harvested" berries are sorted and color-graded, then crushed, frozen, or dried for shipment.

Dried Cranberries are becoming more easily available, but they are treated with vegetable oil and fructose. These berries are less suitable for herbal preparations.

Homegrown Clean and wash the berries in a large sieve. Plunge them into boiling water for thirty seconds to soften their skins and then dip rapidly in cold water. After dipping, put them between sheets of kitchen towels and pat

RIGHT Fresh Cranberries can be frozen and used at a later date.

dry. Lay them out on shallow trays and leave out in the sun to dry. Alternatively, place them in a cool oven heated to 120°F (49°C) for one hour, then increase the temperature to 130°F (54°C). After another hour, increase to 140°F (60°C). Maintain that temperature until the berries are hard and rattle on the trays.

LEFT Harvested berries can be dried or made into preserves.

BOUNCEBERRIES

One-legged New Jersey schoolteacher "Peg Leg" Webb was the first to bounce Cranberries down a flight of stairs to separate the good, firm ones from the soft, squashy ones that never made it to the bottom. Or so the story goes.

Each Cranberry contains four pockets of air, which enable it to "bounce" and also to float. This is why Cranberries are also known as Bounceberries.

Fresh Cranberries

Preparations for internal use

Cranberry capsules

ABOVE
Cranberries make nutritious, healthy drinks.

FOR ALL THESE RECIPES, *the commercial Cranberry is as effective as the fruit from the uncultivated plant. Whether you choose to use commercial Cranberry juice, buy the dried fruit, purchase the prepared capsules, or make your own decoction, tincture, or syrup, they will all be beneficial in their own way.*

CRANBERRY DRINKS

Cranberries make cleansing, tonic, and refreshing drinks. Sweeten them with cold-pressed honey or maple syrup, then drink immediately for maximum nutrition. Try mixing this cocktail: grapefruit, which is rich in vitamins and minerals and strengthens the heart; unpeeled peaches, which are high in vitamins A and C, and minerals; apricots, which are rich in vitamin A;

BELOW *These fruits combine well with Cranberry.*

unpeeled apples, which contain vitamins and minerals and add natural sweetness.

TO MAKE A CRANBERRY DRINK

STANDARD QUANTITY

1lb (450g) fresh Cranberries
5 cups (1.2 liters) spring water
honey to taste

1 *Liquidize the Cranberries, transfer to a pan that is neither copper nor aluminum, cover with the spring water and cook gently for 10–15 minutes.*

2 *Drain through a cheesecloth for a clear juice or through a stainless steel sieve if you do not mind a little sediment. Add honey to taste.*

TINCTURE

For those who cannot drink large quantities of Cranberry juice, this is a good way in which to take advantage of all the nutrients in the fruit. You can keep some of the tincture in a small bottle to carry with you.

The fresh or dried fruit is prepared with alcohol: this extracts the vital nutrients, kills germs, and prevents deterioration. It is most important that only perfect fresh fruit or newly bought dried fruit is used. Decaying or stale berries will have lost much of their goodness and therefore make an inferior tincture.

ABOVE *A tincture seals in the fruit's healing properties.*

TO MAKE A TINCTURE

STANDARD QUANTITY

Use 8oz (225g) of dried Cranberries or 11oz (310g) of fresh Cranberries, chopped into small pieces, added to 4 cups (1 liter) of alcohol and water mixture.

1 *Put the fresh or dried berries into a liquidizer and cover with vodka; 45% proof is standard, but 70–80% proof is better. Liquidize the ingredients – the mixture will be stiff and hard, but persist. Pour the mixture into a large, sterilized, dark glass jar. Cover with an airtight lid. Shake well, label the jar, and store in a dark place.*

2 *After two days, measure the volume of the contents and add proportional amounts of water – add 20% water if using standard vodka and 50–60% water if using 70–80% proof vodka. The whole*

tincturing process will take at least 14 days, but you can leave it for up to four weeks. Shake the jar daily.

3 *Strain the tincture through a fine mesh bag, preferably overnight, until you have the very last drop. For best results you can use a wine press.*

4 *Pour the thick, dark red liquid into sterilized, dark-colored jars and label clearly. Store in a cool, dark place. For personal use, decant into a 2fl oz (50ml) tincture bottle.*

For dosages, see page 34.

RECOMMENDED DOSAGES
FOR TINCTURES

Everyday use *Take approximately 1 tsp (5ml) of tincture three times a day, diluted in approximately 10 tsp (50ml) of water (not fruit juice). It should be noted that practitioners often use double this dose for up to two days during the acute phase of a problem, for example cystitis.*

Long-term use *This herb can be taken for weeks or months if needed. The adult dose is 40 drops two to three times a day.*

Children's dosages *Over 12 years, adult dose of 40 drops 2–3 times daily; 7–12 years, half adult dose; 3–7 years, quarter adult dose; younger than 3 years, use a sweetened decoction instead – up to ½ cup (125ml) daily.*

Periodic use *Some people prefer to take Cranberry for a few weeks and then to have a break for a similar period. This routine can help to keep the brain chemicals balanced without needing to take the herb every day. The dosages are the same as for long-term use.*

BELOW **The whole family can benefit from Cranberry power.**

DECOCTION

This is a delicious way in which to enjoy Cranberries and at the same time to gain benefit from them. Both dried and fresh fruit are equally effective, but the dried fruit produces a sweeter result. Although not as strong as

Cranberry juice, a decoction contains all the same qualities. It may be more suitable for some conditions and for those who wish to avoid the alcohol contained in tinctures.

RIGHT **Dried berry decoctions retain more sweetness.**

TO MAKE A DECOCTION

STANDARD QUANTITY

Use ¾oz (20g) dried Cranberries (presoaked overnight if possible), or 1½oz (40g) fresh Cranberries, chopped or shredded, to 3 cups (750ml) cold water, reduced to about 2 cups (500ml) after simmering.

1 Place the presoaked berries in a saucepan with the water (a double boiler is ideal). Bring to a boil, and then simmer for 20–30 minutes. The liquid should reduce by approximately one third. Remove the pan from the heat.

2 Cool the liquid and strain it into a pitcher. If you are going to store the decoction for longer than a day, keep it in the refrigerator in sterilized, dark glass jars.

For children (and adults) a little sweetening may be needed to soften the tart flavor. Add cold-pressed honey or maple syrup to taste; use it when the decoction is almost cool.

Recommended dosage

Adults: 2 cups (500ml) daily. Children: over 12 years, adult dose; 7–12 years, half adult dose; 3–7 years, quarter adult dose; under 3 years, 3–5 drops twice daily.

BELOW **Simmer gently to extract the berries' essence.**

SYRUP

Cranberry syrup is very palatable and strengthening, and particularly appealing to toddlers and children. It is therefore useful as a gentle remedy for tummy troubles and bed-wetting. Use dried or fresh fruit. Dried berries need to be presoaked for 1–2 days.

TO MAKE CRANBERRY SYRUP

STANDARD QUANTITY

2½lb (1.1kg) of dried Cranberries (presoaked), or 4½lb (2kg) fresh Cranberries, with enough spring water to cover them. You will also need brandy and vegetable glycerine (see step 6 below)

1 Put the berries in a blender, then add enough spring water to cover them by 1in (2.5cm).

2 Process the mixture to break open the berries. Alternatively, crush them with a pestle and mortar. Let the mixture soak for another day.

3 The next day, transfer the mixture to a pan and boil for two minutes, then simmer for 30 minutes. Turn off the heat and let stand for another 30 minutes.

4 Strain off the liquid, and refrigerate when cool. If necessary, squeeze the juice through a fine but strong cheesecloth bag. A wine press is ideal for this job.

5 Now measure the total volume of all the simmered and strained berry liquid. Simmer it on a low heat until it reduces to one quarter of the original volume.

6 Measure the mixture again, then add one quarter of its volume in brandy and vegetable glycerine – for example, 4 cups (1 liter) of Cranberry concentrate will need 1 cup (250ml) of vegetable glycerine and brandy mixed. Use the best brandy you can afford. Pour the syrup into sterilized, dark bottles and refrigerate. It will keep indefinitely.

Recommended dosage

Adults: 6–12 tbsp (90–180ml) a day. Children: over 12 years, adult dose; 7–12 years, half adult dose; 3–7 years, quarter adult dose, younger than 3 years, 2 tsp (10ml) three times daily.

CRANBERRY AND APPLE TEA

One of the most delicious and refreshing teas, an invigorating tonic excellent for convalescents. Although the amount of nutrients extracted from the fruits is not as high as in other preparations, there are enough trace elements present in the tea to have a beneficial effect and to promote vitality.

TO MAKE AN INFUSION

STANDARD QUANTITY

1 tsp (2g) chopped dried Cranberries
1 tsp (2g) dried apple, flaked
1 cup (250ml) boiling water
honey to taste (optional)

LEFT **A teapot infuser makes the whole process very simple.**

1 Put the Cranberries and apple in a tea sock and place it in a cup or teapot. Pour on the boiling water and let it stand for seven minutes.

2 Remove the tea sock and, if desired, add half a teaspoon of organic, cold-pressed honey to the tea (although teas are usually best without added sweeteners).

Recommended dosage
The infusion may be drunk freely by all age groups.

NOTE

Teas can also be made in a special teapot infuser, or in a coffee pot with a plunger.

CRANBERRY AND THYME JELLY

This jelly is a pleasant way to ensure that rich foods are properly digested. It can be served with both savory and sweet dishes. A tablespoon of jelly dissolved in a glass of hot water eases mild stomach troubles and

RIGHT *Cranberry jelly is a treat with turkey pie.*

nausea. Cranberry aids digestion; Thyme helps gastrointestinal healing.

Do not use a copper or aluminum pan: it will destroy the jelly's vitamin C content.

TO MAKE A JELLY

INGREDIENTS

large bunch of fresh Thyme, crushed
9oz (250g) fresh Cranberries, washed
9oz (250g) sharp apples, washed and sliced
water (see step 1)
juice of 1 large lemon
light brown or cane sugar (see step 6)

1 Put the Thyme in a pan with the Cranberries and apples. Add just enough water to cover, bring to a boil, then simmer gently until the fruit is soft. Add a little more water if necessary.

2 Drain overnight through a fine mesh bag into a clean bowl.

3 The next day, measure the juice: for each 2½ cups (600ml), have ready 1lb (450g) of sugar.

4 Return the juice to the pan and heat gently. Add the lemon juice and then the sugar, stirring well until it has dissolved.

5 Bring to a boil, then boil hard until a set is obtained. Since the pectin content of both fruits is high, this should not take long. Transfer into clean, sterile jars. Seal tightly.

CRANBERRY AND APPLE CRUMBLE

One of the best ways to encourage every member of the family to eat Cranberries is to include them in this popular dessert. The flavor is unique, both sweet and sharp, and utterly delicious.

TO MAKE A CRUMBLE

SERVES 4

6 large cooking apples, peeled, cored, and cut into thick slices
water (see step 2)
3oz (85g) of dried Cranberries, chopped
clear honey (see step 3)
4 tbsp (50g) butter or margarine
½ cup (75g) all-purpose whole-wheat flour
1 cup (75g) porridge oats
½ cup (75g) demerara sugar

1 *Preheat the oven to 350°F (180°C). Put the apples in a pan with enough water to cover, and cook until they are just soft but not pulpy. Remove from the heat and add the Cranberries, cover, and let stand for ten minutes.*

2 *Drain the juice and mix it with the minimum of honey to sweeten. Put the apple and Cranberry mixture into an 8in (20cm) lightly greased, ovenproof dish and pour in enough juice to cover the fruit.*

3 *Rub the fat into the flour, then add the oats and two thirds of the sugar. Sprinkle the crumble mixture over the fruit and top with the remaining sugar.*

4 *Bake for 20–25 minutes until it is well-browned. Serve hot.*

Preparations for external use

CRANBERRY IS VALUED MAINLY *for its effects on the internal organs, but it is also a great healer of the skin. Many ancient remedies used infusions of the leaf to heal skin irritation, but the berries were also used to heal persistent skin eruptions caused by chickenpox, pruritus, psoriasis, dermatitis, and shingles.*

ABOVE
Berries were used in the treatment of chickenpox.

TO MAKE A POULTICE

STANDARD QUANTITY

3oz (85g) fresh Cranberries, 2fl oz (50ml) buttermilk or natural yogurt, or 2 tsp (10ml) Slippery Elm powder and 2–4 tbsp (30–60ml) olive oil

1 *Take the fresh berries, scald them in boiling water and dry on a paper towel.*

2 *Put in a blender with enough buttermilk to make a stiff paste. Alternatively, use Slippery Elm powder and olive oil to create a firmer, more adhering poultice. You may need to add more olive oil to enable the blender blades to turn easily. Use the mixture immediately.*

3 *Wash the affected area with warm water. Put the mixture onto a poultice (a large plaster or piece of thin gauze) and press onto the affected area. Secure it in place with a clean cloth or plastic wrap. Leave for 20–60 minutes and wash off with warm water. Apply a gentle ointment such as Calendula or Aloe Vera to speed healing and reduce scarring.*

TO MAKE A VAGINAL DOUCHE

STANDARD QUANTITY

¾oz (20g) dried Cranberries or 1½oz (40g) fresh Cranberries, chopped
3 cups (750ml) cold water, reduced to about 2 cups (500ml) after simmering
4 tbsp (60ml) apple cider vinegar

1 *Liquidize the Cranberries in a blender, then transfer to a pan (a double boiler is ideal). Add the water. Bring to a boil, then simmer on a low heat for 20–30 minutes. The liquid should reduce by about one third. Remove from the heat.*

2 *Let the liquid cool and then strain it into a pitcher. Add the apple cider vinegar.*

Recommended use

Use as normal for a douche. If you are unclear about this, follow the instructions that come with a commercial douche kit. Douche kits are available from some health food stores and pharmacies, and by mail order.

CASE STUDY: CYSTITIS

When Donna visited her medical practitioner not long after she was married and said she was suffering from cystitis, he joked that she was suffering from the "honeymoon illness." Understandably she was a little disconcerted until he explained to her that increased sexual activity, hormonal fluctuation, and birth control pills can all create the conditions in which cystitis may occur. As she was unwilling to take the prescription he suggested, he discussed with her a more holistic approach that included drinking 10fl oz (300ml) of Cranberry juice daily until the condition cleared, and thereafter to make it a regular part of a balanced diet that included plenty of fresh fruit, vegetables, and water, and less coffee and processed food.

Natural medicine for everyone

CRANBERRY HAS WORKED *for centuries to bring health and vitality to those who have depended upon it. Full of natural goodness, it can be used in a variety of ways in which all members of the family, from the very young to the elderly, can enjoy it.*

PREGNANCY

Best used during pregnancy as a food, Cranberry may be added to cereals, fruit, or other salads, or mixed in fruit cocktails. Its high vitamin C content makes it essential for good health during this time, especially in the cold winter months or during exposure to illness. Pregnant women are usually advised not to take antibiotics, so Cranberry juice is a good alternative for bladder infections and cystitis.

However, it is essential to take professional advice before using this or any other natural remedy.

LEFT *Scatter a handful of berries over a leafy salad.*

A drink of Cranberry juice several times a week, in small quantities, or a few dried Cranberries, will protect against constipation as well as diarrhea. However, do not overdo it, because as with most fruit, over-indulgence can actually lead to stomachache or diarrhea.

Cranberries, in moderation and as part of a healthy diet, will ensure strength and vitality for both mother and child.

BELOW *Pregnant women should take Cranberry as a food.*

CHILDREN

Cranberries are particularly good for children to ingest because they help strong and healthy growth of bone and muscle, promote good skin, and reduce the risk of infection. They also minimize the effects of many of the ailments associated with childhood and, most particularly, build resistance to the ever-present winter colds whose breeding ground is the school bus, classroom, and swimming pool. Cranberry drinks can be made palatable and colorful enough to vie in popularity against bottled "pop," while dried Cranberries are a sweet and deliciously tempting alternative to candies.

ELDERLY PEOPLE

Many elderly people suffer from urinary problems, and a small glass of Cranberry juice every day is essential to provide some protection against them when they occur.

BELOW *Give your child invigorating Cranberry drinks.*

The Cranberry fruit will also improve an elderly person's general state of health and build immunity against colds and flu. When it is not possible to exercise very often, Cranberry will relieve toxicity, constipation, and lethargy, which will lead to a stronger sense of well-being.

BELOW *Cranberry strengthens the immune system in the old and young alike.*

Herbal combinations

HERBAL COMBINATIONS *are used to complement the effect of a single herb. However, if you are pregnant, breastfeeding, or have a serious medical condition, you should consult your doctor or qualified herbalist before trying these combinations.*

ABOVE **Cranberry works well on its own or with other herbs.**

Herbal treatments often contain one main ingredient, with others added to produce a healing "rainbow effect." The main herb may be required to soothe impaired tissue, for example, with the others there to nourish, to help eliminate toxins, and to assist in nerve and blood supply. Some formulas have equal quantities of four or five herbs, their similar actions working in slightly different ways.

Marshmallow root

Pau d'arco bark

Dried Cranberries

Corn silks

Uva Ursi leaves

ABOVE **These herbs make an excellent treatment for cystitis.**

CYSTITIS

Made up as a tincture, this combination of herbs makes a complete formula for treating cystitis.

Formula Equal parts of Pau d'arco bark, Uva Ursi leaves, Cranberries, Corn silks, Marshmallow root.

Dosage Adults: 1 tsp (5ml) three times daily for the duration of acute symptoms, and for a further seven days after symptoms have subsided.

Pau d'arco is a general antiviral, antiparasitic, antibacterial, and antifungal herb. So, whatever the cause, obvious or underlying, of the microbial imbalance of cystitis, this herb can rectify it by increasing oxygen supply at a local level.

Uva Ursi is a powerful diuretic renowned for its use against cystitis because of its dual ability to tighten inflamed tissue and reduce infection.

Cranberry's antibacterial, tannin-rich, and anti-inflammatory qualities will reduce tissue inflammation and help clear infection.

Corn silks are very soothing for the pain and irritation of cystitis, as well as being a premier diuretic, thereby allowing water and toxins to be released.

Marshmallow lends a strong demulcent (soothing) element to this formula, which is vital to counter the heat, irritation, and pain of cystitis.

DIARRHEA

This herbal combination, taken as a drink, is simple and yet very effective for any kind of digestive upset where diarrhea is involved – whether caused by bacteria, fungi, or viruses.

Formula
3 parts Cranberry juice, 1 part Aloe Vera juice.

Dosage Adults: 1 cup (250ml) three times daily during acute phase and 1 cup (250ml) once daily after the symptoms have subsided. Children: over 12 years, adult dose; 7–12 years, half adult dose; 3–7 years, quarter adult dose; younger than 3 years, 1 tsp (5ml) once or twice daily.

ABOVE *Aloe Vera is soothing and antiseptic.*

Cranberry is a powerful treatment for diarrhea, as already mentioned in this book.

Aloe Vera leaf juice can be bought commercially. The Aloe Vera plant is renowned for its effectiveness in treating any form of digestive upset that results in diarrhea.

DIGESTION

This combination of herbs for the digestive system is best taken as a tincture. The taste in the mouth is vital to instigate the first process of digestion – saliva production. The taste will be stimulating, slightly hot, and sweet.

Formula Equal parts of Cranberries, Fennel seeds, Gentian roots, Ginger roots, and ½ part of Peppermint leaves.

Dosage Adults: 1 tsp (5ml) six times daily, before and after meals. Children: over 12 years, adult dose; 7–12 years, half adult dose; 3–7 years, quarter adult dose; younger than 3 years, consult a qualified herbalist.

RIGHT (clockwise) *Cranberries, Fennel, Ginger, Peppermint, and Gentian all smooth the digestive process.*

Cranberry helps the digestive process; Fennel is antispasmodic and treats wind, colic, and irritable bowels; Gentian root stimulates saliva production and gastric juice secretions. Ginger reduces nausea, and Peppermint disperses indigestion.

Taken before meals, this formula rallies all the digestive processes from the pancreas to the liver, and stimulates their function. Digestive enzymes will be secreted in a more balanced way ensuring that food is processed effectively and completely, and that undigested food does not linger, causing allergic responses and illness.

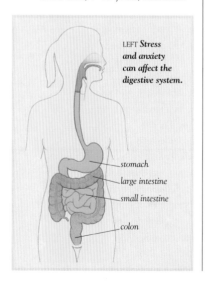

LEFT *Stress and anxiety can affect the digestive system.*

stomach

large intestine

small intestine

colon

PROSTATE INFECTION

Problems with the prostate gland tend to affect men over the age of fifty. This combination of juice and tincture is very easy to administer and the results can be very effective. Simply add Saw Palmetto and Echinacea tinctures to Cranberry juice.

Formula 1 cup (250ml) Cranberry juice, 2 tsp (10ml) Saw Palmetto tincture, 1 tsp (5ml) Echinacea tincture.

Dosage Adults: consume the above quantity three to four times daily during the acute phase and then, after infection appears to have cleared up, reduce to twice daily for another week.

Cranberry's astringent and antimicrobial qualities will soothe and clear the infection rapidly.

Saw Palmetto berries are renowned for their ability to normalize the prostate and alleviate symptoms such as inflammation and pain.

RIGHT **Add tinctures of Saw Palmetto and Echinacea to Cranberry juice to treat prostate infection.**

LEFT *Tinctures must be stored in dark glass bottles to preserve their healing qualities.*

Echinacea will boost the immune system to deal with the infection.

> **CAUTION**
>
> People with autoimmune system problems should consult a doctor or qualified medical herbalist before taking Echinacea.

SKIN-CLEARING FORMULA

The skin responds well to this tincture. It will clear a variety of skin conditions, from psoriasis to heat rashes.

Formula Equal parts of Cranberries, Barberry root bark, Cleavers herb (aerial parts), Burdock root.

Dosage Adults: 1 tsp (5ml) three times daily. Children: over 12 years, adult dose; 7–12 years, half adult dose; 3–7 years, quarter adult dose; younger than 3 years, consult a qualified herbalist.

The high antioxidant and antibacterial content of Cranberries gives the tincture a powerful basis for clearing the skin.

Barberry root bark contains strong antibacterial constituents. It also effectively stimulates bile secretion, the lack of which is at the base of many skin problems.

Cleavers herb is cooling to very hot and inflamed skin conditions. It is strongly diuretic and able to assist in detoxifying the skin.

Burdock is a major cleanser for chronic skin problems because it is able to eliminate the waste products that accumulate in the skin. It is also a powerful natural antibiotic.

Barberry root bark   *Cleavers*

Burdock root

Dried Cranberries

LEFT *Cranberries, Barberry, Cleavers, and Burdock form a powerful skin healer.*

LEFT *Dried Cranberries and Cherry bark, supported by other herbs, keep winter snuffles at bay.*

WINTER CHILLS

The overall effect of this tincture is tasty, sweetly sour, and very soothing. It is ideal for chills, coughs, and colds.

Formula Equal parts of Elderberries, Cranberries, Cherry inner bark, Pine needles.

Dosage Adults: 2 tsp (10ml) three times daily during acute stage, then 1 tsp (5ml) three times daily for 4 days after symptoms have subsided. Children: over 12 years, adult dose; 7–12 years, half adult dose; 3–7 years, quarter adult dose; younger than 3 years, 4 drops every four hours, totaling no more than 16 drops a day.

Elderberries are renowned for their antiviral, vitamin C, and flavonoid-rich components, which are used extensively to treat flu, coughs, and colds.

Cranberries complement Elderberries and have a similar action in this formula.

Cherry bark is a general tonic and specific for loosening phlegm and mucus buildup in the respiratory organs, while at the same time soothing irritated membranes, and easing coughs.

Pine needles are a very powerful antioxidant and are renowned as a wide-ranging antimicrobial rich in volatile essential oils, which help to reduce excessive phlegm, relieve coughs, and improve impaired breathing.

BELOW *The winter chills formula helps keep children healthy.*

How Cranberry works

ALTHOUGH WILD CRANBERRY (Vaccinium oxycoccus) *was traditionally used in remedies by herbalists of another age, it is the more robust* macrocarpon *species that is now used extensively in research into the healing properties of Cranberry.*

LEFT **Cranberries are rich in minerals that fortify the blood.**

The key chemical constituents of the fruit of *Vaccinium oxycoccus* and *macrocarpon* species are very similar, but they are present in the berries in different ratios.

Under normal conditions the urine in the bladder is sterile, but when harmful microorganisms, particularly *E. coli*, proliferate in the bladder, prostate, or kidneys, then a condition exists that is known as urinary tract infection. The high level of concentrated tannins (proanthocyanidin) present in Cranberry juice creates a hostile environment that prevents the adherence of bacteria to mucosal surfaces in the

ABOVE **American Cranberry plants have a high yield of fruit.**

urinary tract. Cranberry juice has also been shown to relieve vaginitis and irritable bladder conditions.

Cranberry contains a high level of vitamin

RIGHT
Vitamin C-rich Cranberries help to ensure clear, healthy skin.

C, which not only ensures the formation of healthy collagen, skin, bones, and muscle, but also encourages normal body cell functioning and promotes the way in which the body uses other nutrients, such as iron, vitamins A and B, calcium, and certain amino acids. By promoting the formation of strong connective tissue, vitamin C increases the healing potential of wounds and burns. It also acts as an antioxidant to protect the body from free-radical damage.

Vitamin C and pectin reduce infection in skin eruptions and irritations. They effectively cleanse and dry

LEFT *The healing powers of the fruit are impressive.*

the area and ensure scar-free healing. Pectin also removes heavy metals, environmental toxins, radioactivity, bacteria, and infection from the body. Vitamin A improves vision and ensures healthy mucous membranes of the urinary, respiratory, and digestive tracts.

Cranberries also contain the minerals calcium, copper, phosphorus, magnesium, and iron, which enrich and purify the blood, leading to higher energy levels.

Cranberry also contains the flavonoids quercetin, myrecetin, and kaenferol, which are vital for protecting blood vessels from damage and lessening the likelihood of disease.

BELOW *Convalescents can regain their strength with Cranberry.*

RESEARCH

For many years, doctors had been somewhat skeptical of the old wives' tale that Cranberry juice is very effective in treating urinary tract infections. Some research had been undertaken in 1914, but it was inconclusive. Research continued for the next eighty years, most of which, yet again, proved inconclusive.

Recently, however, research by Howell and others (see page 58) has moved away from the belief that acidification of the urine was the mechanism through which Cranberry juice produced a bacteriostatic effect (see page 56). It has revealed instead "that proanthocyanidins are the key to the effectiveness of Cranberry in the treatment of urinary tract infections" (see Howell *et al.*, page 58).

As mentioned earlier (see page 23), proanthocyanidins stop bacteria, such as *E. coli*, from adhering to mucosal surfaces in the gut and the bladder. The result is a bacteriostatic effect, whereby the bacteria are not destroyed but do not multiply. In other words, when bacteria are prevented from sticking to the surfaces of the gut or bladder, they cannot proliferate and cause further damage.

Placebo-controlled studies of urinary tract infections in the United States have, since 1994, produced fairly conclusive results. It was found that, of those women who drank 10fl oz (300ml) of Cranberry juice cocktail daily for six months,

ABOVE *Ruby power in a glass: pure Cranberry juice.*

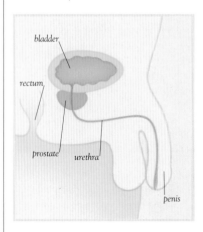

ABOVE *Cranberry juice protects against infections of the urinary tract.*

ABOVE *Research into the beneficial effects of Cranberry continues.*

60% were less likely to develop infections than those who drank a placebo. Among those who already had infections, 75% were more likely to have these clear up.

The key to many of the healing properties of Cranberries are the flavonoids, which inhibit damage from free radicals, those harmful molecules that lead to many serious health problems, including artery disease, heart disease, and cancer. Cranberries contain the flavonoids quercetin, myrecetin, and kaemferol, and it is now considered that the antioxidant properties of flavonoids, and quercetin in particular, help to protect the lining of blood vessels from damage, thereby preventing the onset of diseases of the arteries. Investigation is now underway into the possibility that these compounds may prevent genetic changes that lead to cancer.

Recent research undertaken by Gary D. Stoner of the Ohio State University Comprehensive Cancer Center has shown that ellagic acid, an antioxidant compound, is present in many soft fruits, including Cranberries. Ellagic acid has been shown to inhibit carcinogenic agents and to hinder the growth of tumors. Not only is it powerful enough to help prevent cells that have been exposed to carcinogens from becoming cancerous, but further research has shown that it may also help to prevent mutations in DNA.

RIGHT *These berries may yet be another weapon in the fight against cancer.*

Conditions chart

THIS CHART is a guide to some of the ailments that Cranberry can treat, but it is not intended to replace other forms of treatment. Always consult your doctor or other qualified practitioner before embarking on a course of treatment.

NAME	INTERNAL USE	EXTERNAL USE
ACNE	Juice, tincture	
BED-WETTING	Tincture	
COMMON COLD	Juice, tincture, fresh or dried berries, syrup, decoction	
CANDIDA	Juice, tincture	
CONSTIPATION For adults For children	Dried berries, tincture Syrup	
CHICKENPOX	Decoction	Poultice, decoction used as a wash
CYSTITIS	Juice, decoction	Douche

NAME	INTERNAL USE	EXTERNAL USE
DERMATITIS	Juice, decoction	Poultice, decoction used as a wash
DIARRHEA	Juice, tincture, syrup, decoction	
GOUT	Decoction, tea	
INFLUENZA	Juice, tincture	
PSORIASIS	Tincture, decoction	Poultice, decoction used as a wash
RHEUMATISM	Decoction, tea, juice	
SHINGLES	Tincture, juice	Poultice, decoction used as a wash
SKIN CONDITIONS *Inflamed spots and pustules*	Juice, tincture	Poultice, decoction used as a wash
SUNBURN	Juice	Poultice, decoction used as a wash
URINARY TRACT INFECTION (UTI)	Juice, tincture	Douche
WARTS AND PLANTAR WARTS	Juice, tincture	
WOUNDS AND BURNS	Juice, capsules	Poultice

Glossary

AERIAL PARTS
Those parts of the plant that grow above the ground – stem, leaves, and flowers.

AMINO ACID
Building blocks of protein essential for growth and body maintenance, but which the body does not make.

ANTIBACTERIAL
Destroys bacteria, or arrests bacterial growth and reproduction.

ANTI-INFLAMMATORY
Reduces inflammation.

ANTIMICROBIAL
Destroys or inhibits the growth of microorganisms.

ANTIOXIDANT
A substance that delays or prevents oxidation, which is a normal part of energy production in cells, but which produces potentially harmful free radicals.

ANTISCORBUTIC
Cures or prevents scurvy.

ANTISPASMODIC
A substance that limits, corrects, and prevents excessive involuntary muscular contractions.

ASTRINGENT
In medical terms, this means a substance that draws together or constricts tissue, and acts as a contracting or styptic agent; in cosmetic terms, it means an agent that tones and firms the skin; in the context of taste, it is sharp and acidic.

BACTERIOSTATIC
Stops bacteria from multiplying, but does not destroy them.

CANDIDA
A genus of yeasts that inhabit the vagina and alimentary tract and can, under certain conditions, cause a yeast infection.

COLLAGEN
An insoluble protein, it is the principal fibrous component of connective tissue in the body.

DIURETIC
A substance that increases the volume of urine, and hence the frequency of urination.

FLAVONOIDS
Compounds responsible for a wide range of actions. They can be antiasthmatic as well as anti-inflammatory, antimicrobial, anticarcinogenic, and antioxidant.

FREE RADICALS
Highly reactive particles that damage cell membranes, DNA, and other cellular structures.

GENITOURINARY SYSTEM
This encompasses all the organs and systems connected to the reproductive and urinary systems.

HYPOGLYCEMIA
A deficiency of glucose in the bloodstream, causing muscular weakness and mental confusion.

IMMUNE SYSTEM
Function of the body to fend off foreign bodies, and/or to disarm and eject them.

PATHOGENS
Unfriendly parasitic microorganisms, such as harmful bacteria, that produce disease.

PECTIN
Soluble gumlike carbohydrate formed in fruits from pectose by ripening or heating.

PROANTHOCYANIDINS
Biologically active substances that stop harmful bacteria from adhering to mucosal surfaces in the bladder or gut (see page 52).

PROSTATE GLAND
A male sex gland that opens into the urethra just below the bladder. During ejaculation it secretes an alkaline fluid that forms part of semen. The prostate may become enlarged in elderly men.

PSORIASIS
A chronic skin disease in which scaly pink patches form on the scalp, knees, elbows, and other parts of the body.

TANNINS
Organic substances obtained from plant material.

UTI
Urinary Tract Infection.

Bibliography

Avorn, J., Monane, M., et al., JOURNAL OF THE AMERICAN MEDICAL ASSOCIATION (1994): 271: (10): 751–754

Blatherwick, N.R., ARCHIVES OF INTERNATIONAL MEDICINE (1914): 14: 409–450

THE COMPLETE ILLUSTRATED HOLISTIC HERBAL, David Hoffman (Element Books Limited, 1996)

ENCYCLOPEDIA OF MEDICINAL PLANTS, Andrew Chevallier (Dorling Kindersley, 1996)

THE ESSENTIAL BOOK OF HERBAL MEDICINE, Simon Y. Mills (Arkana, 1993)

FOLK REMEDIES FOR COMMON AILMENTS, Anne McIntyre (Gaia Books Limited, 1994)

THE HEALING GARDEN, David Squire (Contemporary Books, 1998)

THE HERBAL FOR MOTHER AND CHILD, Anne McIntyre, (Element Books, 1992)

Herbalgram web site: www.herbalgram.org

HISTORY OF THE ENGLISH HERB GARDEN, Kay N. Sanecki (Ward Lock, 1992)

Howell, A.B., et al, NEW ENGLAND JOURNAL OF MEDICINE (1998): 339: (15): 1085–1086

HOME HERBAL, Penelope Ody (Dorling Kindersley, 1995)

LEAVES FROM GERARD'S HERBALL, Marcus Woodward (Dover Publications, Inc., 1969)

A MODERN HERBAL, Mrs. M. Grieve (Tiger Books, 1992)

NATURAL FIRST AID, Mark Mayell (Vermilion, 1996)

NIEWE HERBALL, Henry Lyte, 1578, translated from the Flemish of Dodoens and dedicated to Queen Elizabeth I

POTTER'S NEW CYCLOPAEDIA OF BOTANICAL DRUGS AND PREPARATIONS, R. C. Wren (Health Science Press, 1973)

SPIRITUAL PROPERTIES OF HERBS, Gurudas (Cassandra Press, 1988)

THORSONS GUIDE TO MEDICAL HERBALISM, David Hoffmann (Thorsons, 1991)

Zafriri, D., Ofek, I., et al, ANTIMICROBIAL AGENTS AND CHEMOTHERAPY (1989): 33: 92–98

Useful addresses

British Herbal Medicine Association (B.H.M.A.)
Sun House, Church Street, Stroud,
Glos. GL5 1JL, UK
Tel: 011 44 1453–751389
Fax: 011 44 1453–751402
Works with the Medicine Control
Agency to promote high standards of
quality and safety of herbal medicine

Herb Society
Deddington Hill Farm,
Warmington, Banbury,
Oxon OX17 1XB, UK
Tel: 011 44 1295–692000
Fax: 011 44 1295–692004
Educational charity that disseminates
information about herbs and
organizes workshops

Cherry Cyster
Strathenry House, Leslie, Fife
KY6 3HY, UK
Tel: 011 44 1592-620685
Fax: 011 44 1592-620965
For douche bags and other useful
health items.

SUPPLIERS IN THE UK

Baldwin & Company
171–173 Walworth Road,
London SE17 1RW, UK
Tel: 011 44 171–703 5550
Herbs, storage bottles, jars, and
containers available

Hambleden Herbs
Court Farm, Milverton,
Somerset TA4 1NF, UK
Tel: 011 44 1823–401205
Organic herbs by mail order

Herbs, Hands, Healing
The Cabins, Station Warehouse,
Station Road, Pulham Market,
Norfolk IP21 4XF, UK
Tel/fax: 011 44 1379–608201
Organic herbal formulas; Superfood;
mail order and free brochure.

SUPPLIERS/SCHOOLS IN THE USA

American Botanical Pharmacy
PO Box 3027, Santa Monica,
CA 90408, USA
Tel/fax: 1310 453–1987
Manufacturer and distributor of
herbal products; runs training courses

Blessed Herbs
109 Barre Plains Road,
Oakham,
MA 01068, USA
Tel: 1800-489-4372
Dried bulk herbs are available by
mail order in order to make your
own preparations

United Plant Savers
PO Box 420,
E. Barre,
VT 05649, USA
Aims to preserve wild Native
American medicinal plants